Health & safety in
MOTOR VEHICLE REPAIR

HS(G) 67

CONTENTS

7 INTRODUCTION

9 SERVICING AND MECHANICAL REPAIR
- 10 Lifting equipment
- 12 Electrical safety
- 14 Compressed air equipment
- 15 Vehicle inspection pits
- 16 Petrol fires
- 17 Brake and clutch linings
- 18 Wheels and tyres
- 19 Batteries and chargers
- 20 Used engine oils
- 20 Engine running
- 21 Rolling road and brake testing equipment
- 21 Moving and road testing vehicles
- 22 Vehicle valeting
- 23 Steam and water pressure cleaners

25 BODY REPAIR
- 26 Flamecutting and welding
- 28 Noise in body repair
- 28 Vibration
- 29 Body filling and preparation

31 PAINTING
- 32 Storing and mixing paints
- 32 Paint mixing systems
- 33 Paints containing isocyanates
- 34 Paint spraying
- 35 Drying and curing ovens
- 36 DIY spray booths
- 38 Spraying in a single room workshop
- 38 Maintenance and inspection of spray booths and ovens

41 ORGANISING HEALTH AND SAFETY IN MOTOR VEHICLE REPAIR
- 42 Law
- 44 Safety policies
- 45 Accidents and emergencies
- 47 Preventing ill health from exposure to toxic substances
- 48 General working environment
  - *Hygiene and welfare*
  - *Cleanliness*
  - *Floors and gangways*
  - *A safe place to work*
  - *Lighting*
  - *Comfort*

51 FURTHER INFORMATION

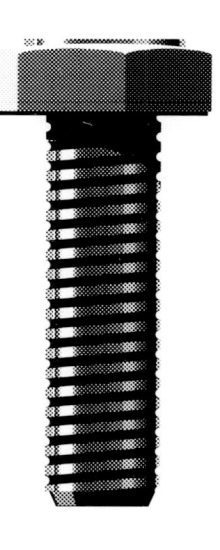

INTRODUCTION

1 There are about 20 000 bodyshops and 36 000 mechanical repair workshops in the United Kingdom. Motor vehicle repair workshops are found in a variety of premises ranging from railway arches and factory units to purpose built main dealerships. However, the fact that motor vehicle repair work is commonplace should not be allowed to obscure the risks involved in some of the activites taking place and the substances being used.

2 Every year over 2000 accidents in garages and vehicle repair workshops are reported to the Health and Safety Executive (HSE) and to local authorities. Many more go unreported.

3 Most accidents involve trips and falls or poor methods of lifting and handling; serious injuries often result from these apparently simple causes. Accidents involving vehicles are very frequent and cause serious injuries and deaths every year. Work on petrol tanks in particular causes serious burns, hundreds of fires and some deaths each year.

4 There is also widespread potential for work related ill health in garages. Many of the substances used require careful storage, handling and control.

5 This booklet is aimed particularly at those managing motor vehicle repair work - employers large and small, and supervisors as well as the self-employed. It will also be of great value to employees. It deals with the more common and specific hazards of mechanical and body repair.

6 The booklet also provides practical advice on how to organise health and safety in garages and explains which laws apply to motor vehicle repair work and how to comply with them. Advice is also provided on how to write a safety policy, how to plan for, control and report accidents and emergencies and what first aid to provide. Six simple check-lists to monitor the general working environment are also provided.

7 The advice in this booklet represents best and good practice. Those who follow it will in most circumstances comply with the law. The chances of anyone suffering an accident or damage to health will also be significantly reduced.

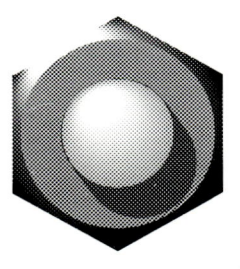

**SERVICING AND
MECHANICAL REPAIR**

LIFTING EQUIPMENT

Hoists

8   Hoists collapse and vehicles fall off them because of failures to maintain and use hoists properly.

9   Four post hoists should have effective "dead mans" controls, toe protection and automatic chocking. Toe traps should also be avoided when body straightening jigs are fitted. Raised platforms should never be used as working areas unless proper working balconies or platforms with barrier rails are provided.

10   Careful attention should be paid to manufacturers' recommendations when using two post hoists - vehicle chassis, sub-frame and jacking points should be in good condition; support arm pads should be set to the correct height before the vehicle is raised. The weight distribution of the vehicle being lifted and the effect of the removal of major components should be constantly evaluated.

Jacks and axle stands

11   Jack vehicles only on level undamaged floors with a trolley or bottle jack rated to lift the vehicle weight safely.

12   Use only jacks maintained in good condition, and only lift from the correct vehicle jacking points.

13   Use the hand brake and/or chocks to stop vehicles moving.

14   Use jacks for lifting the vehicle only; use axle stands in good condition and properly positioned, to support the vehicle weight. Ensure that the correct support pins in good condition are used for the extendible columns of axle stands.

15   Never get beneath a vehicle supported only by a trolley jack or jacks.

Examination of lifting equipment required by law

16 Some lifting equipment is subject to statutory testing examination and certification:

(a) Check that new chains, ropes and lifting tackle have Certificates of Test and Examination specifying safe working loads before they are used - they are normally provided by suppliers or manufacturers.

(b) Have them and hoists and lifts thoroughly examined every 6 months by a competent person (often employed by an insurance company). Obtain and keep the report.

(c) Ensure tests and thorough examinations of cranes are carried out before they are first used and obtain a Certificate of Test and Examination specifying safe working loads. Periodic thorough examinations at least every 14 months are also required for which a report should be obtained and kept.

(d) Do not forget the examination of cranes, lifting and restraining equipment fitted to or used with road recovery vehicles. Table 1 summarises the main legal requirements.

PLANT	SECTION OF FACTORIES ACT	TEST & THOROUGH EXAMINATION PRIOR TO USE	CERTIFICATE OF TEST AND EXAMINATION	PERIODIC THOROUGH EXAMINATION
CHAINS, ROPES & LIFTING TACKLE	26	YES EXCEPT FOR FIBRE ROPE & FIBRE	YES SPECIFYING SAFE WORKING LOAD	AT LEAST EVERY 6 MONTHS
HOISTS & LIFTS	22	NO	NO	AT LEAST EVERY 6 MONTHS
CRANES & OTHER LIFTING MACHINES	27	YES	YES SPECIFYING SAFE WORKING LOAD	AT LEAST EVERY 14 MONTHS

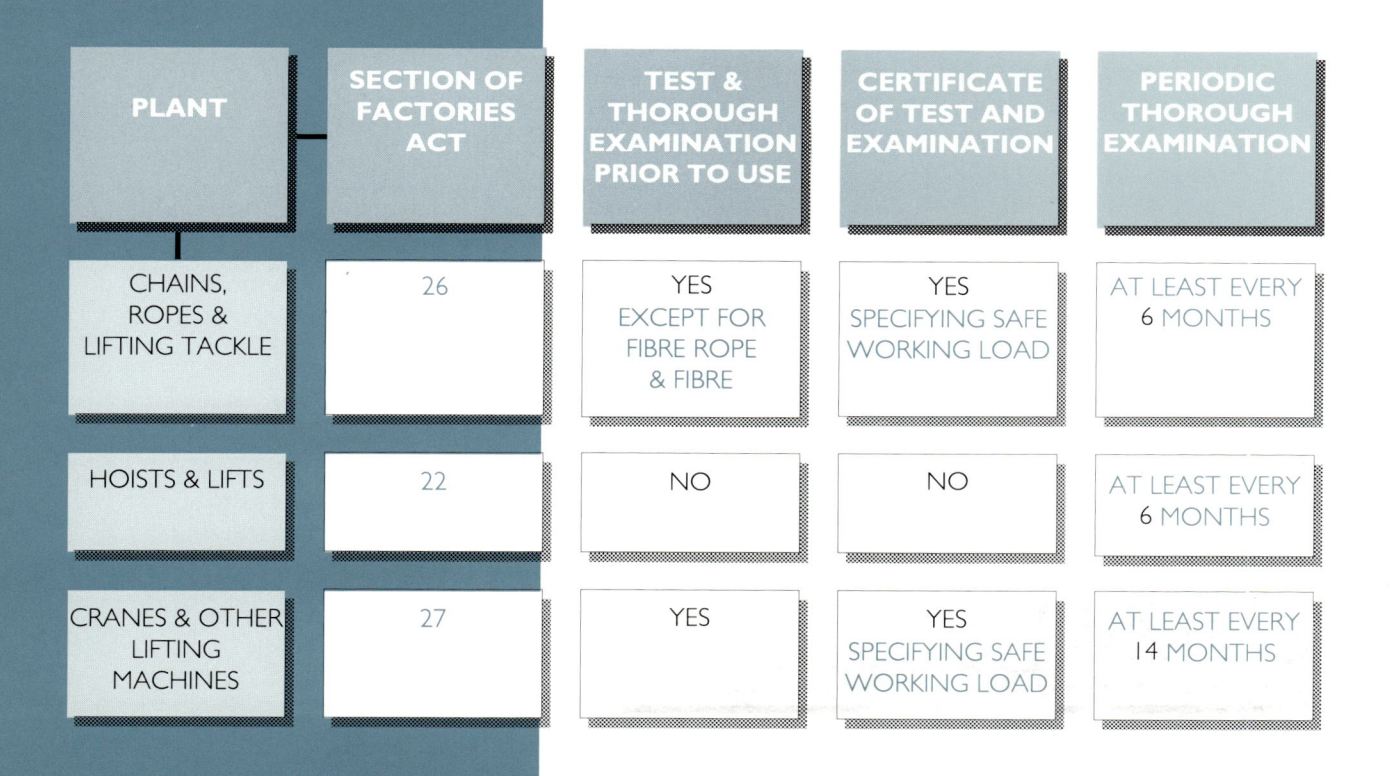

Table 1:
Statutory examination of lifting equipment - Factories Act 1961

ELECTRICAL SAFETY

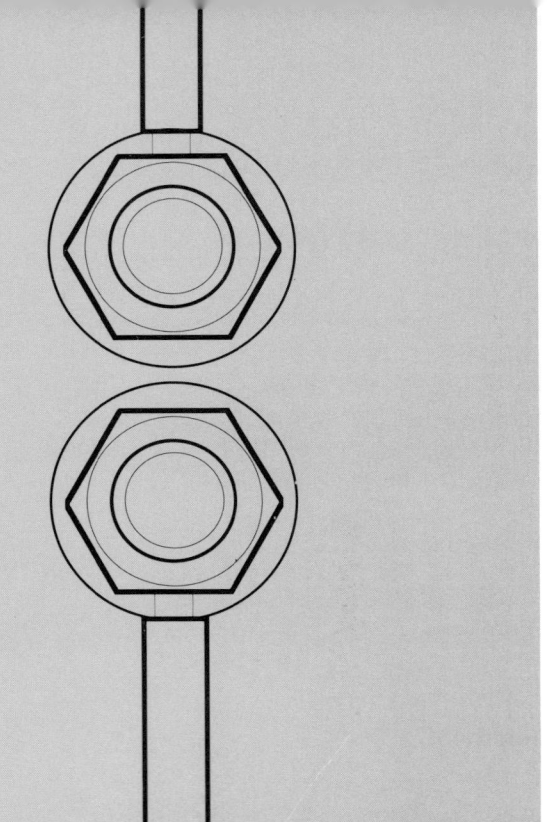

17 Many people are injured in garages by electric shocks caused by poor electrical standards; these also cause many fires and explosions.

Fixed electrical installations

18 Ensure that all equipment is designed for the environment in which it is to be used and suitably protected.

19 Locate switchgear where it won't be damaged but make it readily accessible and unobstructed for repair. Ensure that switchgear suitable for the supply and distribution system is provided.

20 Provide fused switches or circuit breakers at the main switchboard to control supplies to individual circuits and distribution boards.

21 Label all switches and fuseways clearly to indicate the circuit or function controlled. Keep switch and distribution board covers closed.

22 Protect wiring against mechanical damage, preferably by using PVC insulated wires in steel conduit and/or trunking or PVC steel wire armoured cable with an outer PVC sheet. Plastic covered minerally insulated cable may also be used.

23 In workshops all parts of the fixed electrical installation should be one metre above floor level - to remove the risk of igniting spilt petrol or flammable liquids.

24 Provide a generous number of socket outlets on stanchions and walls above bench level to reduce the number and length of trailing leads.

25 Suitable lighting in mechanical repair areas includes pendant type lights with tungsten filaments or fluorescent fittings lamps. Twin fluorescent lamps with phase displacement between lamps will reduce the danger of stroboscopic effects from rotating parts which make moving parts appear stationary.

26 In vehicle washes lights should be totally enclosed and hose-proof. Fix lights to the runways of vehicle lifts where it is necessary to light the underside of vehicles.

27 Only totally enclosed hose-proof type handlamps operating at 24 volts or less from a double wound transformer should be used in the wet.

28 Additional precautions have to be taken in paint spraying and during work on fuel systems to prevent the ignition of flammable vapours. (See paragraphs 55 & 136).

Hand lamps

29 Unsuitable and poorly maintained handlamps cause many electric shocks. They should either be:

    (a) "all insulated" or "double insulated", the bulb protected by a robust cage of insulating material or a transparent insulating enclosure; or

Maintenance of electrical equipment

Avoid danger:

▲ Plan effective and economic maintenance by a competent person.

▲ Use a qualified electrician.

▲ Inspect wiring, installations and equipment periodically.

▲ Inspect portable equipment and flexible leads and connected plugs frequently; weekly checks would not be excessive.

▲ Keep records of inspection and maintenance noting dates of inspections and remedial work carried out.

(b) supplied by reduced voltages such as 110 volts (centre tapped to earth) and SELV which does not exceed 50 volts ac supplied from a double wound transformer which gives electrical separation mains input power or 120 volts dc (ripple free). SELV bulb filaments are heavier and more robust than normal types and are more suited to rough usage..

30   **Low voltage and insulated handlamps offer no protection against the risk of igniting petrol or other flammable vapours and must not be used where such vapours may accumulate especially in vehicle inspection pits and paint spraying areas.**

31   Isolate lamps from supplies before changing bulbs.

32   Ensure cables are properly connected to plugs and lamps - the cable restraint or grip should effectively clamp the sheath of the cable to prevent the cable cores from pulling free of terminal posts.

Portable electrical equipment

33   Portable 240 volt tools and handlamps and their plugs, sockets and flexible leads are often sources of electric shock and burn accidents, some of which are fatal. Air operated hand tools do not pose a risk of electric shock.

34   Use industrial type plugs and sockets to British Standard BS 4343; these are robust, available as drip and weather proof types and are keyway and colour coded; the latter may help prevent low voltage appliances being plugged into higher voltage sockets.

35   Extension leads should be flexible - never use semi-rigid cable of the type used for domestic wiring; neoprene covered cable resists damage from oil. Extension leads with 13 amp fittings should always have an earth wire.

36   Electric portable tools should preferably be operated by 110 volts supplied from socket outlets suitably located and fed from a transformer with the 100 volt secondary output winding centre-tapped to earth so that the maximum possible shock voltage to earth is 55 volts.

37   "Double insulated" or "all insulated" tools are a valuable precaution against electric shock where a 240 volt supply to portable tools has to be used. These however are not suitable for wet environments.

38   Low voltage equipment offers no protection against the risk of flammable vapours being ignited by electrical equipment.

39   The Institution of Electrical Engineers recommends that fixed electrical installations are tested at least every 5 years and that the person doing this test should prepare an inspection certificate for the occupier.

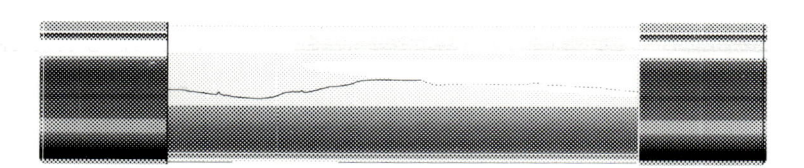

COMPRESSED AIR EQUIPMENT

40   Compressed air equipment is used to power tools and other equipment and also to apply materials such as grease, oil and paint to vehicles.

41   Compressed air equipment, including air receivers, needs to be examined regularly by a competent person (often employed by an insurance company) who will advise on the frequency and type of examination required.

42   If compressed air is used as a source of breathing air for respirators or breathing apparatus, ensure it meets the necessary requirements and fit suitable filters.

43   Injuries, occasionally fatal, can be caused by accidental or deliberate injection of material and/or compressed air, either through the skin or into a body orifice. Internal organs rupture at low pressures in comparison with those of compressed airlines. Ordinary working clothes do not significantly restrict the penetration of compressed air into the body. High pressure paint guns may inject paint at 3000 to 7000 psi (pounds per square inch) and cause serious injuries. Employees should be made aware of the hazards.

44   Care should be taken to avoid accidental injections when using compressed air equipment, particularly in awkward or confined situations such as inside or beneath vehicles, and when clearing or cleaning guns.

45   Horseplay involving people and compressed air equipment should be strictly forbidden.

46   Because the degree of injury is not always immediately apparent, medical advice should always be sought after compressed air penetration occurs or is suspected.

VEHICLE INSPECTION PITS

47   Flammable vapours from petrol, paints and solvents are heavier than air: they accumulate in pits in ignitable and explosive concentrations. Electrical equipment in the pit should be suitable for use in explosive atmospheres; explosion protected equipment is expensive and may be bulky and heavy.

48   Lighting above one metre from the pit floor may be by sealed lights glazed with toughened plastic such as polycarbonate, wire armoured, laminated or toughened glass fitted flush with pit walls to minimise damage from falling objects.

49   Keeping lights clean and surfacing pit walls with glazed white tiles may make handlamps unnecessary; if they are used within pits they should be of an explosion protected type capable of surviving being dropped two metres. Low voltage handlamps give no protection against the risk of igniting flammable vapour.

50   Portable tools used in pits should be air powered or explosion protected. Motors and controls for pumps test and service facilities should be away from pits or explosion protected.

51   Shallow pits not used for repair work which contain rolling road equipment need not have explosion protected electrics.

52   People are also injured when they fall into pits, and not only those unfamiliar with the premises, but also employees who momentarily forget the presence of an unfenced pit. Fence or board pits when not in use. Keep the time they are left uncovered with no vehicle over them to a minimum - during such times use pit lighting and mark pit edges to indicate the hazard. Restrict access to areas where pits may be unfenced.

Precautions

▲ Assess whether petrol needs to be removed. For example, are you welding close to a fuel line? If you are investigating or working on the fuel system itself, is there a possibility that petrol may be spilled?

▲ If there is a foreseeable risk of spillage remove the petrol safely. Work in a level well ventilated area preferably in the open air, from which sources of ignition have been removed. Disconnect the battery. Never remove petrol from a vehicle while it is over an inspection pit.

▲ Always use a fuel retriever as this minimises risks of gross petrol spillages - preferably retrieving from the filler neck, using adaptors and narrow bore hoses to bypass anti-spill devices where necessary. Alternatively retrieve petrol from the opening for the sender unit BUT only if this is accessible safely and located on the top of the petrol tank.

▲ Occasionally it may be necessary to use a fuel retriever with adaptors to retrieve fuel from fuel lines underneath the vehicle - to remove fuel not emptied from the filler neck or sender unit opening. Check your vehicle maintenance and service instructions or contact your fuel retriever and adaptor supplier.

PETROL FIRES

53 Petrol fires in garages are frequent and severe. Inspectors investigate about 20 such fires each year, many of which cause serious burns; occasionally they cause deaths.

54 A large proportion of the fires investigated involved the removal of a sender unit - often located on the side of the petrol tank - without first emptying the tank. Spills also occur when fuel lines are damaged or when fuel systems are being checked. They also happen when petrol is drained into unsuitable containers.

55 Petrol vapour is invisible and heavier than air and will disperse over considerable areas, collecting in pits, sumps and drains. It is easily ignited (sometimes some distance from the spillage) by matches, cigarettes or heaters with naked flames or electric elements, welding gear or electrical equipment.

56 Vapour may also be contained within clothing onto which petrol has been spilled; an attempt to dry clothing using heaters with naked flames or glowing electric elements may ignite vapour in and around the clothes.

Repairing fuel tanks

57 Never attempt hot work on a petrol tank - send it to a specialist.

Advice to employees on how to prevent asbestos dust getting into the air:

- **DON'T** blow dust out of brake drums or clutch housings with an air line.

- **DO** use properly designed drum cleaning equipment which prevents dust escaping; or

- **USE** clean wet rags to clean out drums or housings. Dispose of used rags in a plastic waste bag while still wet.

- **DON'T** grind or drill linings unless the machine has exhaust ventilation or there is a ventilated booth to do the work in.

- **DON'T** use brushes to sweep up dust.

- **DO** use a special (Type H) vacuum cleaner to remove dust.

- **DO** wet dust thoroughly and scrape it up if you haven't got a vacuum.

- **DO** wear the protective clothing, such as overalls, provided by your employer.

- **DON'T** take the protective clothing home. It should be cleaned by your employer.

- **DON'T** use equipment if it is not maintained and checked. Ask to see the test reports for ventilation systems.

BRAKE AND CLUTCH LININGS

58 Some vehicle parts contain asbestos. Working with them can create dust. Breathing this dust is harmful. Cases of asbestos related cancer have been reported in garage workers. Asbestos dust particles are too small to be seen by the naked eye. And the diseases it causes can take years to develop.

59 The dangerous jobs are:

(a) cleaning brake assemblies
(b) cleaning clutch housings
(c) grinding brake linings
(d) sweeping floors

60 Brake and clutch linings and disc pads may contain asbestos. If in doubt assume that they do.

61 Avoid breathing asbestos dust. Prevent dust getting into the air.

WHEELS AND TYRES

Cars

62 Air blasts from the over inflation of car tyres can cause injuries. Contact with rotating wheels during wheel balancing may cause friction burns or other injuries. Risks may also arise from the welding of car wheels to which tyres are fitted.

Ensure that:

(a) Raise and support vehicles safely

(b) Remove the valve core from tyres to be repaired to ensure it is fully deflated; do not deflate tyres by breaking the bead seal.

(c) Use the right tools in good condition for removing wheel nuts and levering tyres off wheels.

(d) Protect against back injuries by using good lifting techniques to lift tyres and wheels from vehicles.

(e) Inflate tyres to the correct pressure; never inflate car tyres above 40 psi (pounds per square inch).

(f) Use a well maintained accurate pressure gauge with at least two metres of air line between gauge and clip on chuck.

(g) Stand clear of tyres during inflation.

(h) Never weld or flame cut a wheel to which a tyre is still fitted.

Commercial vehicles

63 Repairing commercial vehicle and large plant wheels and tyres gives rise to greater and additional hazards from the generally higher pressures involved and the fact that wheels are often constructed of several segments or components. Because wheel replacement is often more expensive and difficult than on cars welded repairs are attempted more frequently.

See HSE booklet HS(G)62 *Health and safety in tyre and exhaust premises.*

When working on the wheels and tyres of commercial vehicles:

- Find out whether the wheels consist of several components.

- Examine the wheel after removing dirt and rust.

- Replace seriously damaged wheels.

- Examine the replacement tyre internally as well as externally; replace dangerously damaged tyres.

- Ensure that the wheel, any associated parts and its tyre are of the correct size and type. Do not force fit wheels, parts of wheels or tyres.

- Use the wheel and tyre supplier's recommended tyre pressure.

- Use an accurate and properly maintained air pressure gauge to control the rate and pressure of inflation. Air-lines fitted with a clip on chuck and at least two to three metres of hose between chuck and control should be used to enable personnel to stand clear, to the side, of a wheel under inflation.

- Use air-line controls with "dead-man's" handle operation.

- Use only air and nitrogen to inflate tyres.

- Inflate tyres in two stages, slowly at first to no more than 15 psi when the wheel and tyre should be checked to ensure a correct fit. Deflate tyre and re-assemble the wheel and tyre if a correct fit is not achieved. Do not use air pressure to force fit wheel and tyres. Complete inflation using either a suitable tyre cage, or underneath a suitable horizontal clamp, or behind a suitable screen or separation wall.

- Use horizontal clamps or enclosures for very large tyres.

- NEVER weld or flame cut wheels to which tyres are fitted. The heat may generate flammable vapour from any oil or lubricating fluid on the inner rim of the wheel which will be confined by the tyre and may reach an explosive mixture. Explosions from the ignition of the vapour are very violent and can kill.

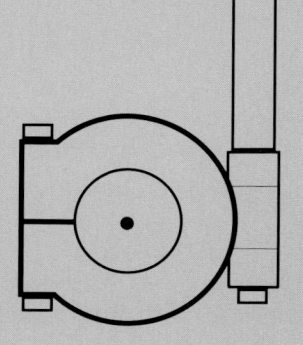

BATTERIES AND CHARGERS

64 During and after charging batteries give off hydrogen, an easily ignited and explosive gas. Connecting or disconnecting batteries or charger connections to battery terminals when batteries are gassing creates incendive sparks. If sparks ignite the hydrogen, batteries may explode spraying those nearby with acidic electrolyte.

65 Do not charge batteries at rates in excess of manufacturers recommendations.

66 Switch off the battery charger before connecting or disconnecting the clips from the charger. If possible connect the clips remote from the battery terminals to for example the starter motor. Keep crocodile clips clean and free from corrosion and, except for the contact surfaces, insulated. Clean battery terminals before fixing charging clips.

67 Do not use battery discharge testers immediately after charging, when false readings are given anyway.

68 If the battery charger transformer is not of the safety isolating type (with an earth screen or windings on separate limbs of the core) one pole of the charging circuit should be earthed - and marked - to prevent a short circuit when the charger is used with a vehicle battery on which a different pole is earthed.

69 Metal in contact with battery terminals causes heavy short circuit currents leading to arcing and/or rapid heating of the metal in contact. Metal finger and wrist jewellery in contact with battery terminals causes burns and flash injuries and should never be worn when working with batteries. Metallic objects should be prevented from falling across terminals wherever batteries are handled, charged or stored.

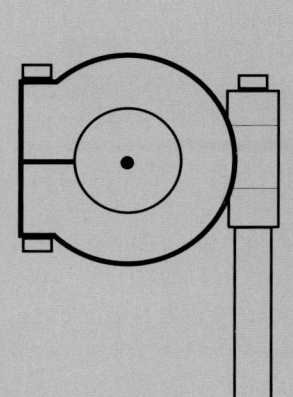

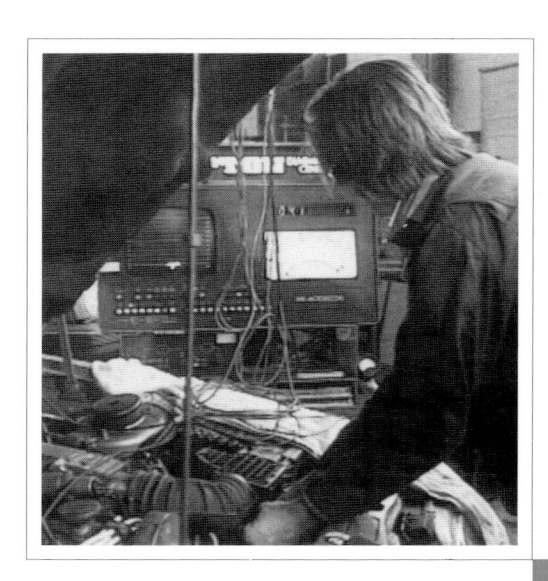

USED ENGINE OILS

70 Frequent and prolonged contact with used engine oils may cause dermatitis and other skin disorders, including skin cancer.

71 Limit exposure to used engine oils by avoiding contact, using safe systems of work and wearing protective clothing which should be cleaned or replaced regularly. High standards of personal hygiene and cleanliness should be maintained.

72 Employees exposed to used engine oils should be encouraged to carry out self-inspection as recommended in HSE leaflet MS(B)5 *Skin cancer caused by oil*. If you have any doubts consult a doctor.

ENGINE RUNNING

73 Exhaust fumes from petrol, diesel and LPG fuelled engines are toxic. They may quickly reach harmful concentrations, particularly from cold or intermittently run engines (when run indoors without exhaust ventilation).

74 Provide extraction or exhaust equipment, preferably by direct coupling to the vehicle exhaust, to ventilate to a safe place in the open air - where fume will not be drawn back into the workshop or affect other premises or people nearby. Maintain couplings and flexible connections in good condition to prevent leaks.

75 Do not rely on catalytic converters to run engines indoors safely. They are less effective when exhaust gases are relatively cool, as from vehicles idling for long periods or used intermittently for short periods. Catalytic converters do not remove toxic oxides of nitrogen.

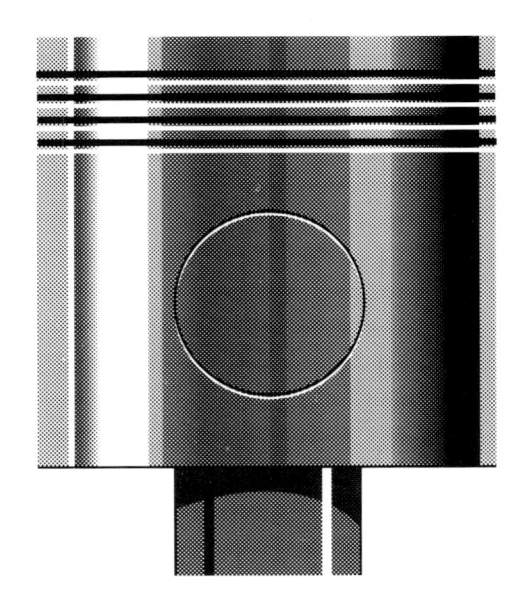

ROLLING ROADS AND BRAKE TESTING EQUIPMENT

76 Serious injuries have been caused by operators attempting adjustments to vehicles under test. To avoid danger:

(a) Ensure "dead mans" controls are fitted and working.

(b) Prevent unauthorised access to areas where testing is carried out.

(c) Fit guards at the sides of rolls where access can't be prevented.

(d) Maintain the running surfaces of the brake tester to prevent unnecessary access for the drying of tyres and the testing surfaces.

(e) Do not carry out other testing or adjustments on the vehicle while the rolling road is moving.

MOVING AND ROAD TESTING VEHICLES

77 Accidents can occur during movement of vehicles on the site and in workshops and when vehicles are road tested. Unqualified, inexperienced and unauthorised drivers are often involved.

Precautions

▲ Make somebody responsible for vehicle movement and testing.

▲ Allow only fully trained and licensed drivers to move vehicles.

▲ Ensure visiting drivers and customers are aware of your rules.

▲ Keep keys secure when vehicles are not in use.

▲ Supervise vehicle movements in restricted spaces, near blind corners and especially when reversing.

▲ Ensure people authorised to drive automatic vehicles are familiar with their operation.

▲ Restrict the testing of high performance cars to older and more skilled drivers.

VEHICLE VALETING

78 Proprietary cleaners often contain toxic and flammable solvents. Concentrations may be high, particularly when used inside vehicles and in poorly ventilated workrooms. Direct skin and eye contact with such solvents is harmful.

79 Adopt the strategy outlined in the Control of Substances Hazardous to Health (COSHH) Regulations to control exposure to hazardous substances. Obtain information from the product label and hazard data sheet. Assess the hazards and risks involved. Where reasonably practicable, use the least hazardous materials available, avoiding materials containing chlorinated solvents, particularly for work inside vehicles.

80 Assess working methods: pour only small amounts of fluid onto a pad or applicator from a small container which is kept closed when not in use.

81 Consider controlling exposure by ventilation:

(a) Ensure the working area is well ventilated.

(b) When working inside vehicles, leave all doors and sun-roofs, if fitted, wide open.

(c) Assess whether the use of local exhaust ventilation with flexible trunking to remove fume and vapour from inside vehicles is reasonably practicable; make sure the fan motor is not in the path of the vapour or is explosion protected.

82 Keep the valeting area free from sources of ignition; disconnect the vehicle battery.

83 Wear protective clothing including eye protection and PVC, natural rubber or preferably nitrile rubber gloves to protect hands and forearms. Remove clothing onto which solvent is splashed and dry in a safe place in the open air.

STEAM AND WATER PRESSURE CLEANERS

84   Fatal accidents can be caused by poorly installed and badly maintained machines often when the lance becomes live because of an electrical fault. Part or all of the machine is often in a wet environment which may increase the potential risk from any electrical fault and also the severity of the shock suffered.

85   Small machines supplied by lower voltages (110 volts) reduce the severity of electrical shocks suffered, and the cables for fixed machines may be less susceptible to physical damage than the flexible cables for mobile machines.

86   Efficient continuity of the earth is vital and circulating current earth monitoring or residual current devices should be used.

87   Establish a system of routine maintenance, testing and repair for installations and safety devices and write it down. Pay particular attention to the plug and flexible cable especially the security and continuity of line neutral and earth connections at each end, and earth continuity bonding between any metal lance, metal reinforced hose and the main earth terminal of the machine.

88   Carry out earth continuity and insulation tests on the machine and on the electrical installation when the equipment is first installed. Regularly carry out simulated fault tests of residual current devices and on circulating current earth monitoring protective devices to prove their reliability.

89   Carry out earth loop impedance tests periodically on supply sockets to prove the effectiveness of the fixed installation earthing conductor in relation to the rating of fuses or other excess current devices which protect the circuit.

90   Instruct and train operators to make external checks each day before use and to report defects immediately. Never use an apparently defective machine.

91   Keep records of examinations, tests and repairs.

92   To prevent **injection injuries** at high pressure steam/water cleaners, never lock the trigger or foot switch in the "on" position. Operators should be trained and instructed in the correct operation of this equipment.

93   Ensure eye protection is worn to safeguard against **eye injuries** from flying debris.

BODY REPAIR

FLAMECUTTING AND WELDING

94 Hazards arise from

(a) the misuse of welding gear and the use of the wrong equipment for the job;

(b) direct contact with heat generated;

(c) electromagnetic radiation;

(d) fires caused by the ignition of flammable material on or near cars such as trim, carpets and upholstery and petrol in tanks fuel lines and nearby containers - often started by sparks or drips of molten metal; and

(e) harmful fumes and gases generated during welding including those from primer and paint layers, other surface coatings such as underseal, and from lead in car bodies.

95 Many of these hazards may be particularly difficult to avoid or prevent when working inside or underneath vehicles.

Arc welding

96 Severe and sometimes fatal electric shocks happen at electric welding apparatus which is designed to operate from mains supply at either 415 volts 3-phase, 415 volts single-phase or 240 volts single-phase. At all installations:

(a) provide fuse protection and mechanically interlock the switch fuse or isolator with the socket outlet so that the plug cannot be inserted or withdrawn with the switch in the "on" position;

(b) earth the workpiece to protect the operator in the event of an interwinding fault between the primary and secondary windings of the transformer. A robust flexible cable terminating in a clamp connected to the workpiece with its other end attached to the metalwork or earth terminal of the power source, is an efficient means of earthing;

(c) during MIG (metal inert gas) welding, prevent contact between the electrode wire and any earthed metalwork to avoid heavy welding current flowing through the earth continuity conductor and destroying it. Use a safe design such as an insulated spool in an insulated chamber in the power source with the wire being fed through insulated rollers and a tube inside the welding cable, to the torch; and

(d) maintain the electrode holder welding current return cables, clamps and safety earths in good condition.

Safeguards for flammable gas cylinders

▲ Store full and empty cylinders in a safe well ventilated place preferably outside buildings.

▲ Never keep cylinders below ground level, next to drains, basements and other low-lying places - heavy gases will not disperse easily.

▲ Some gas cylinders for example acetylene contain liquid - store them with their valves uppermost.

▲ Ensure that workshops where welding is carried out have sufficient high and low level ventilation which is never blocked up to prevent draughts.

▲ Protect cylinders from damage for example by chaining unstable cylinders in racks or on trolleys.

▲ Minimise damage by using the correct hoses, clamps, couplers and regulators for the particular gas and appliance being used.

▲ Turn off cylinder valves at the end of each days work.

▲ Change cylinders away from sources of ignition in a well ventilated place.

▲ Minimise welding flame "flash-back" into hoses or cylinders by training operators in correct lighting up and working procedures and by fitting effective non-return valves and flame arresters.

▲ Use soap or detergent and water solutions - NEVER A FLAME - to test for leaks.

To limit radiation

97 Prevent exposure to direct and reflected ultra violet light and infra red rays by wearing protective clothing, and using welding screens.

98 Use welding screens and wear eye protection to prevent "arc eye".

To control fumes and gases

99 Use local exhaust ventilation wherever possible and always in confined locations. Mobile extraction units with flexible exhaust hoods and trunking can remove fumes and gases from most locations.

100 Be aware of where lead may be used for part of the car body, for example around the tailgates of estate car models, on the headlamp surrounds of older models and as general filler on expensive cars. Use local exhaust ventilation. When welding in these areas supplement local exhaust ventilation with respiratory protective equipment where necessary.

To prevent fires

101 Remove adjacent flammable trim and upholstery before flamecutting or welding especially any onto which molten metal or sparks may fall. Check that fuel lines and tanks will not be affected; empty and remove any which are near or shield them, particularly when welding or flamecutting exhausts.

102 Check that body cavities next to welding or flamecutting are not filled with plastic foam, which is easily ignited; remove where necessary.

To control noise

- get suppliers of machinery and equipment to specify noise levels at operators' positions and choose quiet machines or equipment,

- identify noisy work with "ear protection zone" warning signs,

- isolate bodywork in separate rooms or fix ceiling-high partitions,

- provide information, instruction and training about noise risks to employees, and reduce the duration of exposure by job rotation,

- get operators in noisy areas to wear ear protection,

- as hearing damage builds up gradually, pay particular attention to young people so they can avoid noise exposure before their hearing is permanently damaged.

NOISE IN BODY REPAIR

103 Excessive noise is a serious health hazard. It accelerates the normal hearing loss which occurs as we grow older. Less obvious side effects - increased pulse rate, blood pressure and breathing rate - indicate that noise (and vibration) cause stress.

104 Noise levels are usually measured in decibels - dB(A). A 3 dB(A) increase doubles the noise and the damage it can cause. Noise loud enough to damage your hearing might force you to shout when you speak to someone two metres away. Hearing damage may be caused if noise levels are not reduced, controlled or ear protection provided for workers not worn.

105 Removing and repairing body panels using pneumatic tools is noisy work. Operators using air saws and chisels may be exposed to noise levels as high as 107 dB(A), those using air grinders and orbital sanders 97 dB(A). Noise levels from panel beating and other repair operations using hand tools are variable but generally high; noise from work with sheet metal is often around 93 dB(A). Welding and flamecutting can also be noisy, and paint spraying has been measured at 93 dB(A).

106 Much body repair work therefore will expose operators and others to more than 85 dB(A). Where exposure remains at or above this level throughout the day for any one person, a noise assessment should be carried out by a competent person and ear protection provided. Where daily personal noise exposure exceeds 90 dB(A) take further action to reduce noise, mark ear protection zones, and ensure that all exposed people wear ear protectors. At this noise exposure level employees have a duty to wear the protection provided.

107 The 85 dB(A) and 90 dB(A) action levels are likely to be exceeded where bodywork is a regular daily activity and where pneumatic tools are used even for short periods. Removing panels using an air saw for as little as six minutes can mean that the user's total daily personal noise exposure will exceed 85 dB(A). Using an air sander for 45 minutes can give the user a daily personal noise exposure of more than 90 dB(A).

VIBRATION

108 Direct vibration to the hands from vibrating tools used in body repair can damage bones and joints. A condition known as "vibration white finger" is caused by impaired blood supply to the fingers. Much vibration can be reduced by proper installation, maintenance and use of equipment, although obvious precautions like fitting hand tools with vibration absorbing handles can be taken, a full solution to a vibration problem often requires expert help.

BODY FILLING AND PREPARATION

109 Adopt the strategy outlined in the COSHH Regulations for the assessment and control of the hazards involved in body filling and preparation. Where reasonably practicable, use less harmful materials, exhaust ventilation to control exposure and work with personal protective equipment as a last resort (See publications listed in Further information).

110 Most fillers are reinforced with glass fibre or metal and consist of a thermosetting unsaturated polyester in a solvent (usually styrene) which is hardened by a catalyst. Mixing, applying and finishing such fillers generates toxic fume and dust; the catalyst is often a corrosive irritant and some catalysts are strong skin sensitizers causing dermatitis. (See Dermatitis page 47).

111 Lead is also used in some body preparation operations, and although the temperature at which the alloy is applied is usually not high enough to generate large quantities of harmful fume, subsequent finishing by powered discing and sanding releases high concentrations of fine dust which is a serious health hazard.

112 To minimise the number of people exposed to dust and fume, separate the body filling and preparation area away from other work, preferably in a mechanically ventilated booth fitted with dust tight lighting.

113 Keep dust to a minimum. Large excesses of filler should be removed using coarse hand files. Use powered discing and sanding machines for the final finish only; use tools with built in extraction or local exhaust ventilation.

114 Wear protective clothing, including appropriate respiratory protection. (See publications listed in Further information).

115 When working with lead do not smoke, drink or eat in the workroom. Have separate changing areas for clean and contaminated clothing.

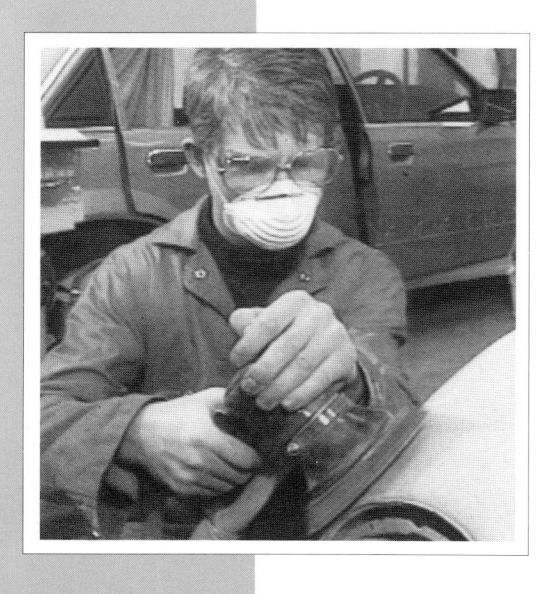

PAINTING

STORING AND MIXING PAINTS

116 Many paints and solvents used in vehicle finishing give off vapour which is readily ignited and which is often toxic. The escape of these vapours should be kept to a minimum. Keep only small quantities (not more than 50 litres) on their own in a metal cupboard or bin for immediate use at the workplace, and larger stocks in a fire resisting store with spillage retention and good ventilation.

117 Keep lids on cans and containers closed to stop vapour escaping. Contain spillages by decanting paint over a tray. Have absorbent material readily available to soak up spillages. Keep contaminated material in a lidded metal bin, and dispose of its contents safely.

118 Exclude sources of ignition and use suitable electrical equipment. Within two metres of paint mixing areas use only explosion protected lighting and electrical equipment. Do not smoke where paints are stored or used.

119 Ensure adequate ventilation where paints are mixed. Breathing protection may be needed.

120 Treat containers emptied of liquid the same as full ones: they will often be full of vapour.

121 Obtain a licence for the storage of petroleum products from your local Petroleum Officer who works for your Fire Brigade.

PAINT MIXING SYSTEMS

122 Proprietary paint mixing systems reduce quantities of paints stored and minimise vapour given off during mixing.

123 Locate the paint mixing unit in a fire resistant well ventilated separate room if possible.

124 Install only explosion protected electrical equipment within one metre of the mixer unit and ensure that electronic balances are protected or a safe distance away.

125 Keep quantities of paints and solvents in the system to a minimum and store stocks of paint and solvents and their empty containers in a separate fire resistant store.

126 Use a work bench with a lip fitted to contain spills.

127 Provide a metal bin for waste rags and materials.

How can health be protected?

- Assess the hazards and risks and prevent or control exposure in line with the COSHH Regulations.

- Identify 2-pack paints from their labels or suppliers' data sheets.

- If you suffer from chronic respiratory disease such as chronic asthma do not work with 2-pack paints containing isocyanates.

- Consult an occupational health professional before working with 2-pack paints for advice on a suitable health surveillance programme.

- Spray only in mechanically ventilated booths or separate workrooms.

- Use only mechanically ventilated ovens for accelerating curing. Run them under negative pressure.

- Ventilate vapour and spray to a safe place in the open air where they will not be drawn back into the workroom or into nearby premises.

- When mixing and spraying wear protective clothing including gloves and eye protection; wear air fed or compressed airline breathing apparatus even for small jobs.

- If a full facepiece canister respirator is worn for mixing or for "touch up" jobs change the canister before its recommended life (often as little as 15 minutes) is exceeded. Gauze facemasks and half masks with cartridges do not provide protection.

PAINTS CONTAINING ISOCYANATES

128 2-pack spray paints containing isocyanates are often used to paint vehicles. In these isocyanate hardeners or activators added to liquid resin and pigments react to produce a polyurethane film.

129 Vapours and spray mists containing isocyanates are highly irritant to the eyes and respiratory tract and may cause asthma. Asthmatic attacks may occur immediately or may be delayed for up to 12 hours after exposure. Symptoms of over exposure are:

(a) sore eyes
(b) running nose
(c) sore throat
(d) coughing
(e) wheezing, tight chest
(f) fever and breathlessness

At first these complaints will generally clear up at weekends or during holidays, but are likely to return when back at work. Some people may become sensitized as a result of working with 2-pack spray paints. In sensitized people, even minute isocyanate concentrations can lead to severe asthmatic attacks. Fatal cases have been reported but these are rare.

PAINTS AND THINNERS

130 Spray and vapour from all types of paints and thinners used in motor vehicle repair are harmful if inhaled. Some paints and thinners are more harmful than others. Assess information from suppliers and choose the least harmful for any particular application. Spraying inside a booth or enclosure should minimise risks to those outside but those inside may therefore be exposed to high concentrations of vapour and spray. When choosing breathing protection consider in detail the job to be done. For example are the insides or undersides of vehicles to be sprayed, where ventilation will be less effective? Different types of breathing protection offer different levels of protection. Compressed airline breathing apparatus with a full facepiece or air fed equipment may be suitable for most spray jobs. Before choosing, assess the work carefully and consult your suppliers of paint and protective equipment. See HS(G)53 *Respiratory Protective Equipment: a practical guide for users*.

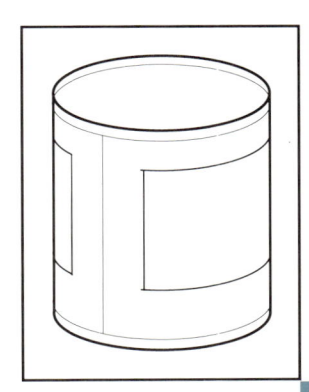

Tips for sprayers

During the refinishing of motor vehicles, the direction, duration and volume of spray is directly under the control of the operator, whose method of work may determine the success of other control measures being applied. Sprayers should:

■ whenever possible stand upstream of the vehicle or part of the vehicle being sprayed. This may be difficult to achieve with side draught booths. Turntables may be used to rotate vehicles. Alternatively down draught booths may be used;

■ not spray above their heads when working in a down draught booth as the spray will be carried downwards by the air flow. When spraying a high sided vehicle in a down draught booth, a suitable platform should be used to make the surface to be sprayed properly accessible and to ensure the correct orientation to the air flow;

■ exercise care when spraying cavities within vehicles. Use lower spray pressures to reduce bounce-back; and

■ when working with another sprayer, work in the same direction and avoid spraying towards one another.

PAINT SPRAYING

131 Spraying gives rise to fine aerosol mists and droplets of toxic and flammable liquid. Vapour concentrations may be high. Safe spraying requires:

(a) effective segregation
(b) adequate ventilation
(c) efficient personal protection
(d) prevention of sources of ignition.

132 To control exposure and contain risks from flammable vapour, spray only in enclosures or booths (proprietary or home made), or in controlled spray spaces.

Proprietary spray booths

133 These are also known as vehicle finishing units and may be separate spray booths or combined with drying or curing ovens. To minimise toxic and flammable risks in spray booths they should be checked and maintained regularly:

(a) Ensure that airflow or air pressure differential switches are working to warn if designed exhaust ventilation flow rates are not maintained.

(b) Maintain any interlocks fitted between spray guns and exhaust ventilation.

(c) Repair damaged spray booth panels to maintain the fire resistance of the unit.

(d) Keep clear escape routes and rescue equipment.

(e) Ensure that air intakes are not obstructed and that discharge vents are correctly sited and in good repair.

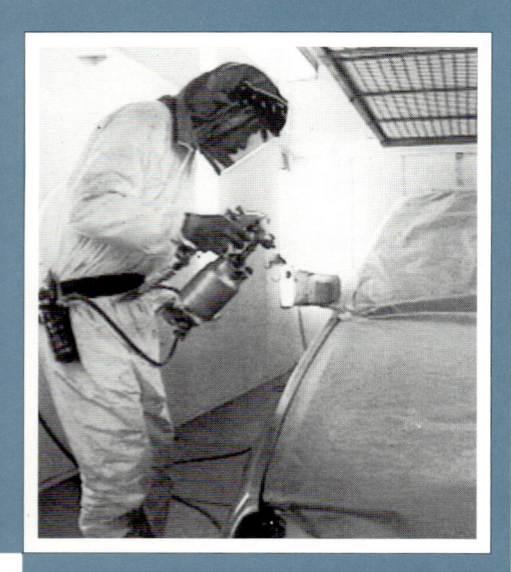

DRYING AND CURING OVENS

134 Drying and curing ovens may be separate or combined with spray booths. Access to them should be restricted when the ovens are working.

135 Check that an adequate exhaust flow rate is maintained by monitoring the position and condition of dampers and the effectiveness of interlocking switches and warning devices.

136 Check that any explosion relief provided is properly maintained; hand pressure should be sufficient to lift the panel. (Explosion relief is required on direct fired ovens, those working at temperatures exceeding 80°C and some ovens in which air is recirculated).

137 For all proprietary spraying and drying booths ensure that sufficient information is received from manufacturers and suppliers on hazards and safe operation to enable booths to be used safely and maintained properly.

Risks from fuel inside vehicles

138 Ventilation equipment and explosion reliefs fitted to spray booths and ovens are not designed to remove risks from fuel evaporating or spilling from vehicles. Fuel vapour pressure rises and risks of fuel and vapour leaks (from petrol, diesel and LPG) increase as temperatures inside ovens rise.

To minimise risks of petrol fires and explosions:

▲ Remove filler caps on all vehicles before they are heated in the unit.

▲ Ensure fuel tanks are reasonably empty (preferably about on quarter full), before they enter the oven; and

▲ Check that fuel lines are intact, particularly at joints.

▲ If the engine has been removed leaving open-ended fuel pipes from fuel tanks in place, the vehicle should not be treated in a vehicle finishing unit oven.

▲ LPG fuel cylinders should be removed from the vehicle before it is put through the drying/baking unit.

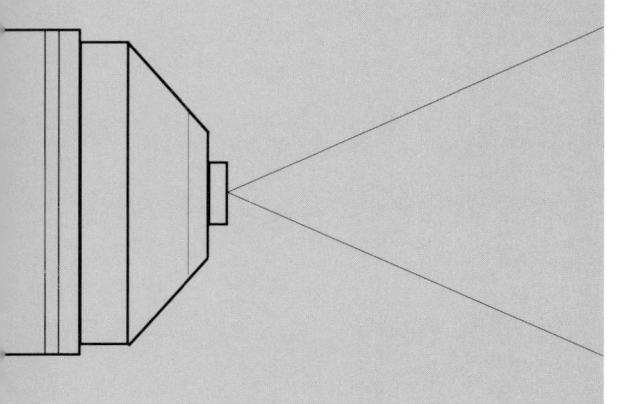

DIY SPRAY BOOTHS

Effective segregation, ventilation and prevention of ignition can be achieved in home-made spray booths, built with professional and competent advice.

Segregation

The separation or isolation should be fire resistant. Some examples of floor, wall and door construction that will provide a standard of half hour fire resistance are .....

Floors

Plain edge boarding on timber joists not less than 38 mm wide with a ceiling of 12.5 mm plasterboard and 12.5 mm of gypsum plaster.

Tongued and grooved boarding not less than 16 mm thick on timber joists not less than 38 mm wide with a ceiling of 12.5 mm minimum of plasterboard and a skim coat of gypsum plaster.

Plain edge boarding on timber joists not less than 38 mm wide with a ceiling of timber lath and plaster, the plaster at least 16 mm thick, covered on the underside with a 12.5 mm thickness of plasterboard.

Walls

100 mm brick (unplastered).

50 mm woodwool slabs plastered at least 12.5 mm thick on both sides, framed construction (non load-bearing).

Steel or timber studding with 12.5 mm portland cement plaster, portland cement/lime plaster or gypsum plaster on metal or timber lathing (non load-bearing conditions only).

Steel or timber studding with 9.5 mm thick plasterboard on each side with the exposed facing of the boarding plastered with 5 mm thick neat gypsum plaster (non load-bearing conditions only).

Where existing walls or partitions are not fire resisting constructions, the standard can be achieved by adding of 12.5 mm plasterboard; ensure that the joints between the overlap are formed over the supporting framework or otherwise suitably constructed.

Doors

Fit: The door should be reasonably straight and true and lie flush against the stop when closed; the gap between the door edge and the frame should not exceed 3 mm.

Door frame: Should have a rebate or stop not less than 25 mm deep; existing planted stops may be replaced or additional material screwed or pinned and glued on.

Door furniture: One pair of metal hinges, all parts of which are non-combustible and have a melting point not less than 800°C.

Glazing: Any plain glazing should be replaced by, or backed with, 6 mm wire reinforced glass not exceeding 1.2 m^2 in area and fitted with solid wood beading not less than 13 mm in cross section.

***Flush doors*:** 6 mm wallboard cover to both sides of the door; fixing to be 32 mm screws at approximately 300 mm centres, or annular nails at approximately 200 mm centres, driven into solid timber.

Panel, framed, ledged and braced doors*: Protection as for flush doors to both faces of the door, or, if protection against fire from one side only, then 9 mm insulating board fixed to room-risk side of the door, as above, with the panels first made up with tightly fitting cutouts of plasterboard or solid wood.

* The importance of fixing cannot be over-emphasised. Additional material must be so fixed to the existing door that, under condition of fire where thermal movement is likely to take place between the door and protective material, the screws or rails are not stressed so that they are pulled out.

Electrics

Unprotected electrical equipment must be kept outside spray booths.

Instal lights outside booths and shine them through fixed and sealed fire resisting wired glass panels.

Use only explosion protected electrical equipment inside the booth.

Heating

Use indirect heating systems such as indirectly fired hot air systems, hot water or low pressure steam radiators.

Ventilation

Get advice from a competent ventilation engineer. Make sure you:

- extract from low level to a safe place in the open air away from people, sources of ignition and nearby buildings and equipment;

- use half hour fire resistant ducting;

- use a centrifugal or bifurcated fan (with the motor outside the ducting in a vapour free area driving the fan through a gas tight shaft seal). Use flexible armoured cable to withstand fan vibration rather than mineral insulated metal sheathed cable;

- use filters to prevent deposits of paint on motor casings, fan blades and inside ducts; deposits may cause fans to vibrate and run out of balance and direct deposits may also cause motors to overheat and ignite;

- ensure fresh air inlets are adequate and provide access points for inspection and cleaning inside ducting; and

- install an outward opening fire escape door in the end wall opposite the booth entrance..

SPRAYING IN A SINGLE ROOM WORKSHOP

139 Where mechanical and body repairs are carried out in the same workshop spray only after:

(a) ensuring adequate half hour fire resistant isolation of the workshop from adjoining rooms;

(b) removing all sources of ignition such as directly fired heaters or domestic-type electric and gas fires and turning off and isolating electrical equipment which is not explosion protected;

(c) providing ventilation to a safe place while spraying; and

(d) ensuring adequate personal protection is worn and that no one else unprotected in the workroom or nearby will be exposed to the spray or vapour.

MAINTENANCE AND INSPECTION OF VENTILATION AT SPRAY BOOTHS AND OVENS AND IN BODY PREPARATION AREAS

140 Examine and test all engineering controls regularly, especially exhaust ventilation.

141 Ventilation in motor vehicle repair is provided to control both flammable and toxic risks. Ventilation in spray booths and ovens prevents the exposure of people outside to fume and vapour given off. It also helps control levels of spray, mist droplets, fume and vapour in the breathing zone of the sprayer.

142 Check spray booths and ovens regularly for leaks; maintain fans and motors. Have spray booths and ovens thoroughly examined and tested by a competent person (either an insurance company engineering surveyor or a representative of the supplier) every 14 months to ensure that control of the exposure of those working outside and inside the booth is maintained.

143 Similarly, in body preparation areas, have local exhaust ventilation for enclosures and tools checked and maintained regularly. A thorough examination and test is also required every 14 months to ensure that the control of exposure of those working outside and inside body preparation areas is maintained.

144 Don't forget to maintain respiratory protective equipment; examine and, when appropriate, test it. Keep written records of equipment tests.

Monitoring ventilation rates

Side draught booths

145 HSE Guidance Note EH 9 on spraying of highly flammable liquids recommends a minimum air velocity of 0.7 m/s at the working opening, but this may need to be increased to 1.5 m/s if particularly toxic materials are used. The figures should be interpreted to be the minimum mean velocities at the booth face.

146 For open fronted booths where vehicles are sprayed measuring points should be regularly spaced at three heights with a minimum of one point horizontally for every metre in width of the booth.

147 Where the operator works inside the booth, the plane of measurement should be where he or she works. The mean of the measurements should not be less than 0.5 m/s with a minimum measured value of 0.4 m/s.

Down draught booths

148 Air speeds should be measured at 10 points around the vehicle, three on each side and two at each end, at 0.5 m from the vehicle and at a height of 0.9 m. The mean of these 10 values should be greater than or equal to 0.4 m/s with a minimum measured value of 0.3 m/s.

149 For long booths used to spray heavy goods vehicles measurements may be made at heights of 1.5 m, 0.5 metres from the vehicle, two at each end and others at intervals between 1.5 m and 2 m along the sides. A mean value of not less than 0.4 m/s and a minimum measured value of 0.3 m/s should be achieved.

Maximise the efficiency of the booth

- Provide means to indicate when dry filters need replacement. The air speed in the immediate vicinity of the sprayer in a dry filter spray booth is often the lowest in the booth because of the accumulation of spray deposits on the filter. Hence the air speed tends to be slowest where it is most needed.

- Keep unnecessary equipment out of booths. Large drums of paint for example can cause recirculation of contaminated air into the sprayer's breathing zone.

- Provide sufficient and suitable lighting in the booth to remove the temptation to spray outside it.

- Give training in techniques of spray painting to teach how to spray with the minimum amount of overspray and bounceback, to obtain the correct balance between air and liquid flow rates, and to ensure that the minimum pressure for good atomisation is always used.

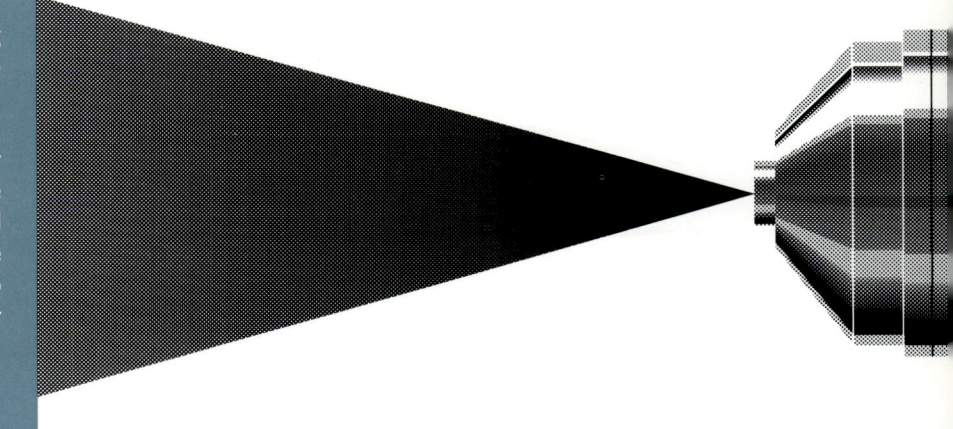

ORGANISING HEALTH AND SAFETY IN MOTOR VEHICLE REPAIR

LAW

Health & Safety at Work etc Act 1974 (HSW Act)

150 If you run a motor vehicle repair business either as an employer or as a self-employed person, the HSW Act requires you to ensure, so far as is reasonably practicable, the health and safety of yourself and others who may be affected by what you do or fail to do.

151 You have duties towards people who:

(a) work for you, including casual workers, part-timers, trainees and sub-contractors (if you are an employer);

(b) use workplaces you provide (if you are a landlord);

(c) are allowed to use your equipment (if you allow friends to repair their own vehicles on your premises);

(d) visit your premises (customers, for example);

(e) may be affected by your work (your neighbours, the public and other workpeople).

152 You also have a duty to yourself, to take reasonable care and to cooperate with others in complying with their duties.

153 The HSW Act applies to all work premises and activities and everyone at work (employee, supervisor, manager, director or self-employed) has responsibilities under the Act. Those who design, manufacture and supply equipment and substances also have duties to see that what they supply is safe, so far as is reasonably practicable.

154 In addition there are specific laws relevant to motor vehicle repair work where people are employed. The main ones are:

(a) the Factories Act 1961; and

(b) the Offices, Shops & Railway Premises Act 1963.

155 Regulations made under these acts set detailed standards for cleanliness, machinery safety and the inspection and examination of certain plant.

156 The Control of Substances Hazardous to Health (COSHH) Regulations 1988 set a legal framework for the control of hazardous substances for employers, employees and the self-employed.

Meeting legal standards

157 Some acts or regulations are specific - setting out, for instance, how flammable liquids should be stored or how a grinding machine should be guarded.

158 Requirements under the HSW Act are usually supplemented by published codes of practice and although you do not have to follow them precisely, you do have to meet their standards in an equally satisfactory way.

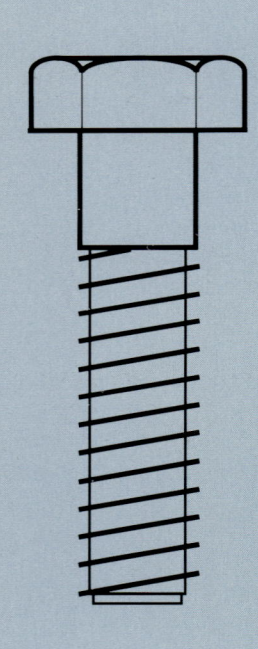

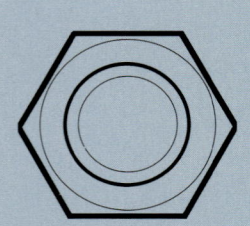

Setting up the business

159 Employers running motor vehicle repair firms must notify their HSE inspector that they are in business before they occupy new premises. They must display certain documents, eg an Employers' Liability (Compulsory Insurance) Certificate and placard copies of certain laws.

160 Children (under school attendance age) are generally prohibited from working in motor vehicle repair work and there are also restrictions on the part-time employment of children in other work such as car washing and working in shops attached to garages.

161 People employed on hazardous processes such as isocyanate paint spraying should be medically examined by an occupational health doctor or nurse as part of a continuing programme of health surveillance. The Employment Medical Advisory Service (EMAS) can help employers contact suitably trained doctors or nurses who can do this work.

162 Body repair work such as paint spraying or body preparation requires protective clothing or equipment to be obtained and provided before starting work.

163 Safety representatives may be appointed by a recognised trade union. They can investigate accidents and potential hazards, pursue employee complaints and carry out inspections of the workplace. They are also entitled to certain information and to paid time off to train for their health and safety role.

164 Someone needs to be trained to carry out certain specified tasks such as mounting abrasive wheels.

165 Competent persons such as insurance company engineering surveyors should be appointed to carry out periodic tests and examinations of equipment such as hoists and air receivers.

Enforcing the law

166 Health and safety laws relating to your firm will be enforced by a factory inspector from the HSE or an environmental health officer from your local council.

167 Inspectors may visit workplaces without notice but you are entitled to see their identification before they come in. They may want to investigate an accident or complaint, or inspect the safety, health and welfare aspects of your business.

168 They have the right to talk to employees and safety representatives, take photographs and samples, and even in certain cases to impound dangerous equipment. They are entitled to cooperation and answers to questions.

169 Inspectors are aware of the special risks of motor vehicle repair work and will give you help and advice on how to comply with the law. If there is a problem they may issue a formal notice requiring improvements or, where serious danger exists, one which prohibits the use of a process or equipment.

170 Inspectors have powers to prosecute a firm (or an individual) for breaking health and safety law.

SAFETY POLICIES

171 Those in charge of every motor vehicle repair business employing five or more people must write down their policy for ensuring health and safety and show it to an inspector if asked.

172 Writing a safety policy statement helps employers put hazards and legal requirements into perspective and to decide on priorities. It should outline the health and safety objectives of the business, the organisation for ensuring the objectives are met and make clear the arrangements for carrying out the policy.

173 For example, personal protective equipment will probably be required for some activities. The safety policy would need to consider:

(a) what is needed;

(b) who is going to issue it;

(c) who will train and instruct people in its use; and

(d) who will check its effectiveness and replace it when necessary?

ACCIDENTS AND EMERGENCIES

174 You need to respond quickly in an emergency, whether it's a simple accident or a major incident. Plan how to deal with possible problems. Some need to be reported to the local office of your inspector.

Controlling an incident

175 Plan for reasonably foreseeable incidents: accidents, broken bones, electrocutions, fires, explosions and chemical spills.

176 Tell people:

(a) what might happen and how the alarm will be raised;

(b) what to do, including how to call the emergency services;

(c) where to go to reach safety or get rescue equipment;

(d) who will control the incident, and the names of other key people such as the first aiders;

(e) essential actions such as emergency plant shutdown or making processes safe.

After an accident or serious incident

177 Treat any injuries and deal with the immediate emergency and make the premises safe.

178 If applicable, report the accident or incident details by telephone and in writing to your inspector. Record any injuries in your accident book.

179 As far as possible during rescue, clearing up operations or your own investigations, take care not to destroy any evidence which will show the inspector what the cause was. If in doubt, check with the inspector.

First aid

180 Immediate and proper examination and treatment of injuries may save life - and is essential to reduce pain and help injured people make a quick recovery. Neglect or incorrect treatment of an apparently trivial injury may lead to infection and ill health. An appropriate level of first aid treatment should be available in the workplace.

181 Appoint someone to take charge in an emergency, to call an ambulance and to look after the first aid equipment. At least one "appointed person" must be available at all times when people are at work.

182 Provide (and keep clean) a first aid box containing only first aid material. The box should contain guidance on the treatment of injured people - in particular how to keep someone alive by artificial respiration, how to control bleeding and how to deal with an unconscious patient. Keep the box near washing facilities.

183 Display notices giving the locations of first aid equipment and the name and location of the appointed person or first aider.

Plan for action

- keep any access ways for emergency services and all escape routes clear;

- clearly label important items like shut off valves, electrical isolators and fire fighting equipment;

- make sure emergency plans cover times when people may be working on their own, such as weekends or when the premises are closed;

- train everyone in emergency procedures such as fire drills, and don't forget the special needs of people with disabilities;

- check emergency equipment regularly - for example the fire fighting equipment;

- help the emergency services by, for example, drawing up a simple plan marked with the location of hazardous items such as paint and gas cylinder stores;

- have a system to account for staff and visitors in the event of an evacuation.

184 In larger repair businesses you may need a first aid room, a qualified first aider or someone with specialist first aid training.

185 First aiders must have training appropriate to the hazards of the workplace. A qualified first aider is issued with a certificate which is valid for three years only - after that a refresher course and re-examination is necessary. Organisations carrying out the training of first aiders must be registered through EMAS - ask your local Employment Medical Advisory Service.

186 In all vehicle repair businesses it makes sense to have someone who knows the basics of first aid, eg resuscitation, control of bleeding, and treatment of an unconscious patient. Make sure your first aid arrangements cover those who work away from their base such as vehicle recovery operators.

Reporting accidents, incidents and disease

187 All injuries should be recorded in an accident book. Some injuries, diseases and dangerous occurrences must also be reported to your inspector.

188 Regulations applying to all employers and the self-employed, and covering everyone at work, require you to:

(a) report immediately by phone to the local office of the HSE or local authority (whichever is responsible for your premises) if as a result of or in connection with your work.....someone dies, receives a major injury, or is seriously affected by, for example, electric shock or poisoning;

(b) there is a "dangerous occurrence" (near miss), including the collapse of lifts and hoists, and explosions or fires which prevent the workplace from being used for 24 hours or more;

(c) confirm, in writing, within seven days, a telephone report of a death, major injury or dangerous occurrence;

(d) notify, in writing, within seven days of the accident, any injury which stops someone doing their normal job for more than three days;

(e) report certain diseases suffered by workers who do specified types of work including occupational asthma arising from exposure to isocyanate and vibration white finger.

189 Have copies of the report form (F2508) ready for use.

Dermatitis

Dermatitis is caused by exposure to chemicals, abrasives, ultra violet radiation and heat - all common in motor vehicle repair. Many oils and greases and solvents such as paraffin, trichloroethylene and white spirit are irritant; resins and their hardeners may cause skin allergy.

Do not under-estimate the extent of the harm caused by dermatitis.

▲ Limit contact with known skin irritants and allergens by using safe working practices or by changing processes.

▲ Use gloves which are impermeable to the materials concerned and avoid contaminating the insides of gloves.

▲ Wash gloves and other protective items - especially before removal - to prevent the spread of contamination.

▲ Wash the skin when contamination does occur.

▲ Use reconditioning creams (and sometimes also barrier creams).

▲ Seek medical advice if the skin becomes red and flaky or begins to blister or crack; rashes may be easily controlled with treatment.

▲ Dermatitis often starts at the site of a minor cut or graze so employers should provide prompt and effective first aid treatment for these injuries.

PREVENTING ILL HEALTH FROM EXPOSURE TO TOXIC SUBSTANCES

190 Many potentially harmful substances are used in motor vehicle repair. It is important to deal with them in the right way. A framework for action is set down by law - the Control of Substances Hazardous to Health (COSHH) Regulations. Lead and asbestos are subject to separate regulations; these however follow the same principles of good occupational hygiene.

191 COSHH requires:

(a) Assessment - of the potential of a substance to cause harm, the hazard, and of the likelihood of harm in actual circumstances of use, the risk. Substances that are hazardous to health include substances labelled as dangerous (that is very toxic, toxic, harmful, irritant or corrosive) under other statutory requirements and substances with occupational exposure limits. In motor vehicle repair pay particular attention to:

(i) paint and thinners;
(ii) cleaning solvents;
(iii) body fillers and hardeners; and
(iv) used engine oils.

(b) Prevention or control of exposure - by using less harmful substances or by controlling exposure by partial enclosure and extraction equipment if reasonably practicable, and by using personal protection where other control measures are not adequate.

(c) Proper use of control measures

(d) Maintenance, examination and test of control measures

(e) Informing employees - employees have to be informed about the risks arising from their work and the precautions to be taken.

(f) You may also need to monitor your employees' exposure and provide health surveillance.

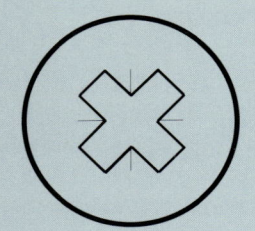

GENERAL WORKING ENVIRONMENT

192 Use the following six checklists to find out what facilities you may need to make your workshops safe and healthy and to provide a reasonable standard of welfare for workers and visitors alike.

Hygiene and welfare

- Separate toilets for each sex (subject to certain exemptions) marked appropriately;
- toilets ventilated, kept clean, in working order and easily accessible;
- ventilated space between toilet and any workroom;
- wash basin with hot and cold (or warm) running water;
- soap and towels (or electric hand dryer) - nail brush where required;
- barrier cream, skin cleansers and skin conditioning cream provided where necessary;
- waste bins (regularly emptied);
- adequate provision for workers away from base;
- drying space for wet clothes;
- lockers or hanging space for work/home clothing;
- clean drinking water supply - clearly marked;
- adequate facilities for taking food and drink, with wash-up sink and means of heating water.

Cleanliness

- Premises, plant and equipment kept clean, with safety information clearly marked;
- good housekeeping to clear trade waste, dirt and refuse regularly;
- rubbish and food waste covered and regularly removed to keep premises clear of pests;
- regular cleaning up of spillages;
- floors and steps washed or swept regularly;
- internal walls and ceilings washed or painted regularly.

Floors and gangways

- Kept clean, dry and not slippery;
- good drainage in wet processes, particularly vehicle washing areas with suitable footwear or working platforms provided where necessary;

- ramps kept dry and with non-skid surfaces;
- gangways and roadways well marked and kept clear;
- level, even surfaces without holes or broken boards;
- floor load capacities posted in lofts, and spares storage areas etc;
- salting/sanding and sweeping of outdoor routes during icy or frosty conditions;
- steps, corners and fixed obstacles clearly marked, eg by black and yellow diagonal stripes.

A safe place to work

- Adequate space for easy movement and access to vehicles;
- no tripping hazards such as trailing wires etc;
- handholds or guardrails where people might fall from floor edges;
- no glass, except "safety glass" in spring doors and on busy gangways;
- neat and tidy storage of tools and equipment;
- furniture placed so that sharp corners don't present a hazard to passers-by.

Lighting

- Good general illumination with no glare;
- regular cleaning and maintenance of lights and windows;
- local lighting for dangerous processes and to reduce eye strain and fatigue;
- no flickering from fluorescent tubes (it can be dangerous with rotating engine parts which may appear stationary when they are not);
- adequate emergency lighting;
- specially constructed fittings for wet, flammable or explosive atmospheres during paint mixing and spraying, underseal application, or vehicle washing;
- outside areas satisfactorily lit for work and access during hours of darkness - for security as well as safety;
- light coloured wall finishes to improve brightness, or darker colours to reduce arc welding flash, for example.

Comfort

- Comfortable working temperature;

- suitable clothing for the job, adequate where necessary for work outdoors and in the wet such as vehicle washing or rescue services;

- good ventilation while avoiding draughts;

- mechanical ventilation where fresh air supply is insufficient;

- an easily read thermometer in the workroom;

- heating systems which do not give off fumes into the workplace;

- noise levels controlled to reduce nuisance as well as damage to health;

- heat stress reduced by controlling radiant heat (especially near head level) and local "hot spots", arising, for example, from paint drying lamps.

FURTHER INFORMATION

Publications

Free leaflets (marked*) and information about HSE priced publications are available from HSE public enquiry points open 10 am to 3 pm Monday to Friday: See inside front cover for addresses.

A complete list of HSE publications, *Publications in series*, is published twice yearly, and is available free from the public enquiry points.

The following are all HSE publications published by HMSO unless other wise marked.

Most priced HSE publications are available from HMSO bookshops, or on request from any good bookseller.

SERVICING AND MECHANICAL REPAIR

Lifting equipment

British Standard BS AU 161 Vehicle lifts. Part 1a:1983 - *Specification for fixed vehicle lifts. Part 2: 1989 - Specification for mobile vehicle lifts*

British Standard BS AU 154a: 1970 - *Specification for hydraulic trolley jacks*

Electrical safety

GN:GS 27	*Protection against electric shock* ISBN 0 11 883583 1
GN:GS 37	*Flexible leads, plugs, sockets etc* ISBN 0 11 883519 X
GN:PM 32	*The safe use of portable electrical apparatus (electrical safety)* ISBN 0 11 883563 7
GN:PM 37	*Electrical installations in motor vehicle repair premises and amendment sheet* ISBN 0 11 883569 6
GN:PM 38 Rev	*Selection and use of electric handlamps* ISBN 0 11 883582 3
HS(G)22	*Electrical apparatus for use in potentially explosive atmospheres* ISBN 0 11 883746 X
HS(R)25	*Memorandum of guidance on the Electricity at Work Regulations 1989* ISBN 0 11 883963 2
*IND(G)89(L) Rev	*Guidance for small businesses on electricity at work*

British Standard BS 4533: Section 102.8: 1990: *Luminaires: Specification for handlamps*

Petrol fires

*IND(G)50(L) Rev   *Safe use of petrol in garages*

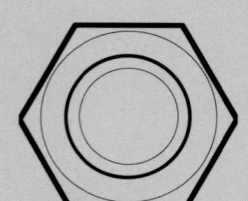

Compressed air equipment

HS(G)39 Rev   *Compressed air safety* ISBN 0 11 885582 4

BS4667:Part 3:1974   *Fresh air hose and compressed air line breathing apparatus*

Brake and clutch linings

*IND(S)14(P)   *Asbestos alert: beware garage dust (poster)*

*IND(G)17(L)Rev   *Asbestos and you*

COP 21   *Approved code of practice: the control of asbestos at work* ISBN 0 11 883984 5

GN:EH 41   *Respiratory protective equipment for use against asbestos* ISBN 0 11 883512 2

GN:MS 13 Rev   *Asbestos* ISBN 0 11 885402 X

Wheels and tyres

HS(G)62   *Health and safety in tyre and exhaust fitting premises* ISBN 0 11 885594 8

SHW428   *Inflation of tyres and removal of wheels (cautionary notice)* ISBN 0 11 880856 7

Moving and road testing vehicles

*IND(G)22(L)   *Danger! Transport at work (factories)*

HS(G)6   *Safety in working with lift trucks* ISBN 0 11 883284 0

COP 26   *Rider operated lift trucks: operator training; approved caode of practice* ISBN 0 11 885459 3

GN:PM 28   *Working platforms on fork-lift trucks* ISBN 0 11 883392 8

Vehicle valeting

GN:EH 5   *Trichloroethylene: health and safety precautions* ISBN 0 11 883606 4

Steam and water pressure cleaner

GN:PM 24   *Electrical hazards from steam/water pressure cleaners* ISBN 0 11 883538 6

F2250   *Pottery (health and welfare) Special Regulations 1950) cautionary notice: hydrofluoric acid* ISBN 0 11 881226 2

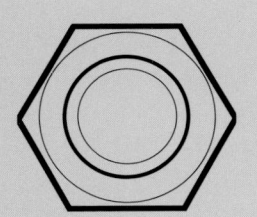

BODY REPAIR

Flamecutting and welding

GN:PM 64 — *Electrical safety in arc welding* ISBN 0 11 883938 1

GN:MS 15 — *Welding* ISBN 0 11 883184 4

Booklet No 50 — *Welding and flamecutting using compressed gases* ISBN 0 11 883080 5

Noise

Booklet — *Guidance on Regulations: Noise at work (Noise Guides 1-2):The Noise at Work Regulations 1989* ISBN 0 11 885512 3

L3 — *Noise assessment, information and control (Noise Guides 3-8) (formerly HS(G)56)* ISBN 0 11 885430 5

Booklet — *100 Practical applications of noise reduction methods* ISBN 0 11 883691 9

*IND(G)75(L) — *Introducing the Noise at Work Regulations*

*IND(G)99(L) — *Noise at work - leaflet for employees*

Body filling and preparation

GN:EH 26 — *Occupational skin diseases: health and safety precautions* ISBN 0 11 883374 X

GN:EH 40 — *Occupational exposure limits (updated annually)* ISBN 0 11 883940 3

GN:EH 44 — *Dust in the workplace: general principles of protection* ISBN 0 11 883598 X

HS(G)37 — *An introduction to local exhaust ventilation* ISBN 0 11 883954 3

COP 2 — *Control of lead at work; approved code of practice* ISBN 0 11 883780 X

*MS(A)1 Rev — *Lead and you*

*MS(B)6 — *Save your skin: occupational contact dermatitis*

SHW366 — *Dermatitis from synthetic resins* ISBN 0 11 880854 0

HS(G) 53 — *Respiratory protective equipment: A practical guide for users.* ISBN 0 11 885522 0

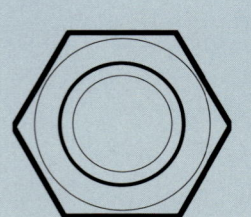

Painting

HS(G)51	*Storage of flammable liquids in containers* ISBN 0 11 885533 6
Booklet	*Guides to the Fire Precautions Act 1971 No. 2 Factories* ISBN 0 11 340444 1
GN:MS 8	*Isocyanates: medical surveillance* ISBN 0 11 883565 3
GN:EH 9	*Spraying of highly flammable liquids* ISBN 0 11 883034 1
GN:EH 16	*Isocyanates: toxic hazards and precautions* ISBN 0 11 883581 5
GN:PM 25	*Vehicle finishing units: fire and explosion hazards* ISBN 0 11 883382 0
HS(G)53	*Respiratory protective equipment: a practical guide for users* ISBN 011 885522 0
Booklet	*Respiratory protective equipment: legislative requirements and lists of HSE approved standards and type approved equipment* ISBN 0 11 885428 3

ORGANISING HEALTH AND SAFETY

Law

*HSC 2	HSW Act: The Act outlined
*HSC 3	HSW Act: Advice to employers
*HSC 4	HSW Act: Advice to the self- employed
*HSC 5	HSW Act: Advice to employees
*HSC 7	*Regulations, approved codes of practice and guidance literature*
*HSC 8	*Safety committees: guidance to employers whose employees are not members of recognised independent trade unions*
*HSC 9	*Time off for the training of safety representatives*
*HSC 11	HSW Act: Your obligations to non-employees
*HSE 4	*Short guide to the Employers' Liability (Compulsory Insurance) Act 1969*
*HSE 16	*Don't wait until an inspector calls ...: the law on health and safety at work: essential facts for small businesses and the self-employed*
L1	*A Guide to the HSW Act* (Formerly HS(R)6) ISBN 0 11 885555 7

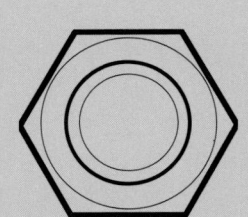

*IND(G)14(L)	*Securing compliance with health and safety legislation at work: how it is done and how it affects you*
*IND(G) 74(L)	*Need advice on occupational health? A practical guide for employers.*

Safety policies

*HSC 6	*Writing a safety policy statement - advice to employers*
Booklet	*Writing your health and safety policy statement. Guide to preparing a safety policy statement for a small business* ISBN 0 11 885510 7

Accidents and emergencies

*IND(G)3(L)	*First aid provision in small workplaces: your questions answered*
*HSE 11	*Reporting an injury or a dangerous occurrence*
*HSE 17	*Reporting a case of disease*
HS(R)11	*First aid at work* ISBN 0 11 883446 0
HS(R)23	*Guide to Reporting of Injuries, Diseases and Dangerous Occurrences Regulations 1985* ISBN 0 11 883858 X
COP 42	*First aid at work: Health and Safety (First Aid) Regulations 1981 and guidance* ISBN 0 11 885536 0

GENERAL WORKING ENVIRONMENT

*IND(G)32(L)	*Watch your step*
*IND(G)33(L)	*Do your signs comply? The Safety Signs Regulations 1980*
HS(G)38	*Lighting at work* ISBN 0 11 883964 0
HS(R)7	*A guide to the Safety Signs Regulations 1980* ISBN 0 11 883415 0
Booklet	*Watch your step: Prevention of tripping, slipping and falling accidents at work* ISBN 0 11 883782 6
GN:EH 22	*Ventilation of workplaces* ISBN 0 11 885403 8

Controlling hazardous substances

*ASP 3	*Your health poster: Dirt and disease*
*MS(A)1Rev	*Lead and you*

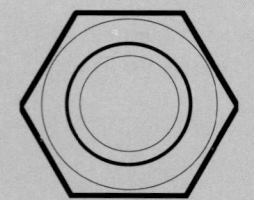

*MS(A)7	*Cadmium and you*
*MS(B)5	*Skin cancer caused by oil*
*MS(B)6 Rev	*Save your skin: occupational contact dermatitis*
*MS(B)9	*Save your skin: advice to employers*
*MS(B)12	*Save your health campaign against occupational lung disease - advice for employees*
*MS(B)16	*Save your breath: occupational lung disease - guide for employers*
*IND(G)60(L)	*Down with dust*
*IND(G)64(L)	*Introducing assessment*
*IND(G)67(L)	*Hazard and risk explained*
*IND(G)72(L)	*Health hazards to painters*
L5	*Approved codes of practice: control of substances hazardous to health* ISBN 0 11 885468 2 (formerly COP 29)
COP 2	*Control of lead at work* ISBN 0 11 883780 X
*HSE 5	*An introduction to the Employment Medical Advisory Service*
HS(G)37	*An introduction to local exhaust ventilation* ISBN 0 11 883954 3
Booklet	*Respiratory protective equipment: legislative requirements and lists of HSE approved standards and type approved equipment* ISBN 011 885428 3
Leaflet	Department of Employment (PL811) *AIDS acquired immune deficiency syndrome and employment* (available from The Mailing House, Leeland Road, London W13 9HL)
SHW366	*Dermatitis from synthetic resins* (cautionary notice) ISBN 0 11 880854 0
SHW367	*Dermatitis* (cautionary notice) ISBN 0 11 880849 4
SHW397	*Effects of mineral oil on the skin* (cautionary notice) ISBN 0 11 883086 4
GN:EH18	*Toxic substances: a precautionary policy* ISBN 0 11 883178 X
GN:EH 26	*Occupational skin diseases: health and safety precautions* ISBN 0 11 883374 X
Booklet	*A guide to health and safety in GRP fabrication* ISBN 0 7176 0294 X

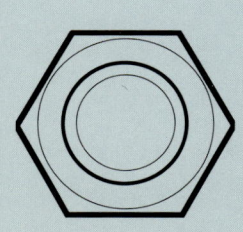

Printed in the UK by HSE and published by HMSO   C200   9/91